Biometrics Fingerprint Attendance Management in an Organization

CONTENTS

Part A: Abstract

IT Problem: A huge business workforce has led to challenges in attendance management and security. Keeping track of who has accessed the premises of organizations or specific classified areas has became challenging because of the enormous traffic flow. Workers have practiced buddy punching, which has led to organizations incurring significant losses due to the salary being paid not reflecting the number of hours worked.

Solution: Fingerprint identification is a common biometric form of the biometric identification system. This project illustrated another sincere contribution to developing, designing, and installing a fingerprint-based identity management system. The primary purpose of implementing the project was to keep track of the daily attendance of workers through fingerprints. Biometrics fingerprint management is an innovative technology that can be implemented to solve the challenge of security and attendance management. The primary goal was to automate the traditional method of marking attendance and generating reports. This system was not only beneficial in keeping track of workers' attendance records but also the time they actually reported to work. Furthermore, it provided an account of who accessed a specific department or section at a given time. This unique identification method prevented errors and impersonation in computing attendance. *The key points of the project implementation, from design to completion:* The project was implemented by first collecting stakeholder views, by assessing an organization's topography,

3

reviewing the current attendance management system, preparing a proposal, designing a biometrics fingerprint system, implementing the plan, pilot phase, training, and post-implementation monitoring. Businesses will carry out this project using the Extreme Programming (XP) technique. This technique is an offshoot of Agile software development methodology that was developed as a standalone desktop application. Microsoft SQL and Microsoft visual basic were used to develop the database and software for building the graphic interface for workers. The system software was tested by deploying it on a laptop device. The system testing was important to ensure that all hardware components, such as the fingerprint scanner, were functioning properly. Experts performed tests to ensure regular and proper synchronization with the database server.

Key metrics used to manage quality and measure the proposed and actual outcomes of the project: The key metrics used for measuring performance included improved worker attendance, reduced vandalism and theft cases, and reduced losses due to truancy and output levels.

Keywords: Attendance management, biometrics, biometrics fingerprint, stakeholders, resource allocation, project management

Part B: Quality Assurance Summary

Biometrics Fingerprint Attendance Management in an Organization

Problem and Its Causes

Attendance management can be a big challenge for an organization, especially when the firm has a vast workforce. For big companies, this problem of attendance is coupled with security challenges. These organizations have to deal with the issue of security that accompanies a considerable workforce. It is challenging to keep track of who has accessed the organization's premises or specific classified areas because of the enormous traffic flow (Adewole et al., 2015). Without an organized program or framework to track entries into the organization and various departments, workers or intruders access unauthorized rooms where they may tamper with machines or equipment that run critical operations in the company. For example, company secrets have often found their way out due to flawed security systems. Some organizations have reported being hacked and their confidential data stolen. This flawed system can motivate rival organizations to manipulate workers within the organization or use them as enablers to gain sensitive data that can destroy the company. When workers are not well-monitored or if they discover a flaw in the system, they are most likely to exploit that loophole, which leads to losses in the organization. Poor attendance implies reduced output per worker, which affects the organization's income because the labor an employee offers does not reflect the salary they are paid to deliver. It equally prolongs project delivery time, which can be a turn-off for many clients. When reliability is poor, customers are likely to seek services elsewhere. Therefore, this exposes the organization to losses and risks such as theft and infiltration by adversaries.

Many organizations have consistently experienced this challenge of attendance management for the past decade. According to an internal audit, over 50% of workers escaped duty severally without official leave in 2021. This trend was dangerous for the organization where the project was trialed because it led to reduced output. Workers practiced buddy punching, making the organization incur significant losses because the salary paid did not reflect the number of hours they had worked. With workers skipping duty but still needing to be paid, it implies that the company has paid them for work not done. For instance, in 2022, the company recorded over 30,000 US dollars in losses, which they attributed to workers' output not reflecting human resource costs. Attendance problems slow down work in an organization. This is because work that should be completed within a short period of time takes longer due to the number of workers assigned to the task being affected by absenteeism. This significantly affects delivery time and leads to dissatisfaction from customers. An unsatisfied customer can result in a bad public reputation for the company's services. Subsequently, this leads to a loss of customers and potential clients, resulting in low profits. Therefore, for the project, this problem informed the company's decision to seek alternative means of managing attendance.

The organization's system of attendance management was time-consuming and labor-intensive. The organization relied on a manual-based attendance management system. This manual-based system required a lot of time to fill and assess, with workers queuing every morning in the central registry to mark their attendance. Because the process of filing attendance was manual-based, workers had to deal with a lot of paperwork. It made the organization incur an extra human resource cost by hiring additional workers to deal with this extra paperwork. The organization also had to employ supplementary clerks to organize and maintain these records. Assessing this attendance was problematic and took a lot of time because of the voluminous paperwork involved (Olagunju et al., 2018). It made senior leaders lazy in reviewing attendance records to ascertain whether workers respected the company's code of conduct in regards to duty. This massive volume of data meant that storing records was equally problematic, and the attendance aspect also became challenging to review when conducting performance appraisals due to the amount of data involved. Equally, it strained the organization's leadership. Senior-level managers had difficulty running the organization with absentee workers. This habit of workers clocking in for each other was a problem for senior management staff. They always found it difficult to keep workers in check because of a flawed attendance management system. Therefore, this prompted the organization to seek external help to curb this attendance management menace.

These manual records were equally prone to distortion, theft, or misplacement. Keeping and maintaining manual records for a long time was difficult because they were prone to distortion by various factors such as water, rodents, or people (Olagunju et al., 2018). This complicated the matrix during retrieval because some files went missing. Human beings can be manipulated to operate to the whims of someone else, especially when money is involved. Workers under investigation in regards to their attendance could deliberately interfere with the process by engineering theft, where they bribed officials in charge to report a file as missing. This is why this approach to attendance management has been proven to be ineffective. Therefore, this prompted companies to seek alternative methods to solve the issue.

As an organization that dealt with technology and graphics, people least expected it to face challenges that technology can solve. In contrast to expectations, the company had incurred costs previously due to replacing items vandalised by unknown individuals. In some instances, it was even forced to buy new equipment because of theft. For example, in 2022, it had to replace over 12 computers due to breakdowns that unknown individuals caused. The accountability system for company property was dysfunctional. Keeping an account of its properties was challenging because management could not track who entered and left the organization. Therefore, corrupt workers exploited this loophole to steal company property for personal use. Not all workers had integrity; hence, when some employees noticed a flaw within the system, they exploited it to their advantage instead of reporting it to management for suitable solutions. A continuation of this trend was dangerous for the organization because competitors could use it to bring the company down when they realized it had flawed security systems. Thus, this universal issue prompted corporations to seek alternative methods of managing security and attendance management.

Biometrics fingerprint management is an innovative technology that companies can implement to solve the challenge of attendance management. This system is beneficial in keeping track of workers' attendance records, the time they report to work, and giving an account of who accesses a specific department or section at a given time. Any attempts of buddy punching or clocking in for each other is easily detectable under this system, as an individual's biometrics must first be used before a record is marked as present (McKenna & Sarage et al., 2015). Equally, individuals cannot access any room without using their biometrics that match the details in the system.

Formative Evaluation

Approach to Formative Evaluation

Experts often use impostors and genuine attempts to evaluate how accurately a biometric system functions. Experts in corporations have assessed the system's biometric performance by using genuine and impostor attempts and then saving all the similarity scores. They used a varying score threshold against similarity scores to calculate the pairs of FAR and FRR (or FMR and FNMR). Subsequently, the experts presented the results either in pairs, that is, in plots or FRR at a certain level of FAR. The experts then expressed the rates in various ways. For instance, in percentages (2%), as fractions (2/100), by using powers often (102), or in decimal format (0.01).

The FAR and FRR are critical when comparing the two systems. The more accurate system often shows lower FRR at the same FAR level. Nevertheless, some systems only report the match/nonmatch decision instead of the similarity score. In such a scenario, there is a higher possibility of gaining a single FRR/FAR pair (and not a continuous series) in the performance assessment result (El-Abed & Charrier, 2012). An adjustable mode of operation (security level) is better because it allows experts to run the performance evaluation repeatedly in different modes to get more FRR/FAR pairs. An adjustable mode of operation implies that experts have a way of controlling the internally used score threshold.

Experts have plotted performance evaluation results using two common methods. First, by using DET graph (Detection Error Trade-off) plots where FRR occupied the Y-axis while the FAR took the X-axis. This graph is all about the false negative vs. false positive rate. A logarithmic scale is often used, especially on the FAR axis (El-Abed & Charrier, 2012). When the Y-axis indicates the number of match errors, the curve near the bottom of the plot represents the greatest biometric performance. Second, experts could equally represent performance evaluation results using ROC graphs (Receiver Operating Characteristic). This graph plots true positive (1 FRR) vs. false positive rate (FAR). Unlike the DET graph, this method shows the best performance biometric close to the top of the plot.

Experts prefer DET curves best because they are far better at showing the critical operating levels and areas of interest.

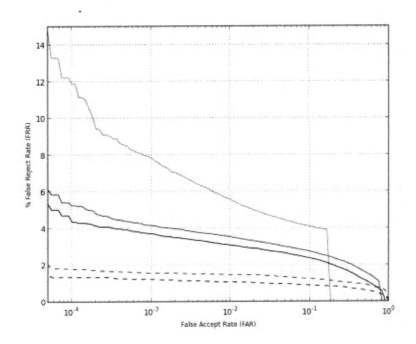

Figure 1: **Example of DET graph**

The red curve at the top represents the worst biometric performance. It shows that FRR of about 8% at FAR 1/1000.

Meaning of Acronyms

FMR – False Match Rate

An FMR is the number of impostor attempts that have been falsely stated to correspond to a template of a different object.

FNMR – False Nonmatch Rate

They represent the number of genuine attempts that are falsely stated not to match a template of the same object.

FTA – FailuretoAcquire Rate

They represent the number of attempts when the system does not produce a sample of acceptable quality.

FAR – False Accept Rate and

FRR – False Reject Rate

Approximately similar to FMR and FNMR, respectively. However, the definitions differentiate between transactions and attempts. A transaction may entail a series of attempts, and contingent on the system's configuration, the result of individual attempts influences the transaction differently. FAR and FRR equally consider the FailuretoAcquire. When a transaction contains precisely one attempt, FRR and FAR are computed as follows:

FAR = FMR * (1 FTA)

FRR = FTA + FNMR * (1 FTA)

EER – Equal Error Rate

The point where the number of False Matches is equal to False Nonmatches (FNMR = FMR).

Test Cases and Scenarios Used

Three ways of performance evaluations exist primarily: Scenario, technology, and operational evaluation.

When evaluating biometric algorithms, experts prefer technology evaluations because they are the most feasible and common. The results of this performance assessment method are reproducible because it is done using saved samples. Equally, this method is less complicated and time-consuming.

1. Technology evaluation

This method conducts evaluations using saved data, for example, previously obtained images of fingerprints.

2. Scenario evaluation

This evaluation method uses a simulated or prototyped environment to assess an end-to-end system.

3. Operational evaluation

This evaluation method is used to assess the performance of a full biometric system in a certain application environment using a specific population.

Experts have used technology, scenario, and operational evaluations to assess the quality of the biometric systems in their companies.

Manipulating Evaluations

The errors observed in these performance evaluations (False Accepts and False Rejects) for the project were not evenly distributed among the volunteers in most instances. Notably, this uneven distribution was caused by the difficulty of capturing some biometric features or with some having minimal unique characteristics. In the fingerprint biometric system, some fingerprints had:

- Dry skin (for instance, due to cold weather).

- Worn friction ridges (for instance, due to manual labor).

- Skin disease.

Some that lacked unique characteristics were due to:

- Failure to present the appropriate area of the finger, for instance, only screening the top of fingers with less characteristic structures.

- Few minutiae in the ridge pattern.

The experts controlled this to improve the calculated biometric performance by selectively removing the worst samples from the database. The risk of manipulation was avoided by self-evaluating the system rather than depending on fabricated numbers. Further, the company hired independent entities that could be trusted to conduct the evaluations, but the conditions and test environments were determined by somebody else.

Explanation of the Acceptance Criteria

Biometric performance was conducted by performing numerous impostor and genuine comparisons and analyzing the generated match decisions or similarity scores. The error rates were computed as a proportion of:

- Impostor attempts that are falsely accepted (FAR).

- Genuine attempts that are falsely rejected (FRR).

FRR at fixed levels of FAR is employed by comparing systems with each other to establish if a system is sufficiently accurate for a particular use case. Experts at the company used the FRR at fixed levels of FAR to determine if a system was accurate for every employee identification process. The samples used in the technology evaluations were then stored in the databases.

The features of a database significantly influenced the achieved biometric performance. Because these experts did not compare the evaluation results obtained from various databases and:

- Equally, they ensured they collected samples that better mimicked the target individual, such as the kind of volunteers, sensors used, and physical conditions. These quality samples ensured that more reliable predictions were made from the assessments.

Statistical confidence in the calculated figures was considered, and people with biometric difficulties were omitted from formative evaluations.

Part C: Detailed Review of the Completed Project

Assumptions

The project's primary assumptions were that the system could succeed in eliminating time theft within the organization and create a seamless management process by eradicating the use of paperwork in attendance management, while improving security, and reducing costs associated with attendance management. The implementation of the system and subsequent trial phases indicated that the fingerprint biometrics system was an efficient method for monitoring workers.

First, a post-assessment of the project indicated that time theft had significantly reduced in the organization after the implementation. The number of workers who could skip jobs for personal business had reduced significantly. Those who skipped job duties had to do so after prior permission from the human resource office. This reduction resulted in positive implications for the organization. First, the time taken for project delivery was reduced immensely. Before this system's implementation, the time for customer project delivery was lengthy because of non-corporations among team members due to consistent absenteeism. This lengthy delivery time was a turn-off for many clients who could later switch to rival companies. However, after this fingerprint biometrics system was implemented, workers were often present and punctual for their duties. Furthermore, the reduction in delivery time ensured that the organization was able to retain its current customers and it even served as an attraction for new clients.

Equally, the organization managed to get value out of its human resource costs due to the elimination of time theft. The time, labor, and output from individual workers accurately reflected the individual's payslip amount. When people work less than the stipulated time, they are stealing from the organization because they are not providing value for the money they are paid. As discussed, this time theft is a big blow to the organization because it scares away customers and leads to financial losses. Therefore, the project fulfilled this first assumption.

Second, a post-assessment of the project showed that the system created a seamless management process because the organization's use of paperwork for attendance management had reduced significantly. One major problem before the project was that the organization spent significant resources on the paper-oriented attendance management system. The organization had to incur extra costs for labor because more workers were needed to help take and review attendance. Reviewing all the paper-based records was problematic because of the volume of data and the tendency for distortion, theft, and loss. Notably, this complicated the matrix in performance appraisals for individual workers and the generation of payslips based on worker attendance. Equally, workers wasted a lot of time every morning and evening by having to manually sign in and out. Before the system's installation, stakeholders assumed it could eradicate these problems and create a seamless organizational management process. Managers could easily conduct performance appraisals using electronically generated reports, review worker attendance without major hurdles, and reduce human resource costs on attendance management. Therefore, this system's installation fulfilled the assumption that it could introduce a seamless management process.

Third, a post-assessment of the project indicated that introducing this system improved security within the organization. One of the problems the company experienced before was vandalism and theft of property, with difficulty in tracking responsible individuals. Stakeholders assumed the system could introduce a way of tracking an individual who entered every room within the building. Consequently, this biometric system simplified the process of tracking potential suspects of vandalism and theft. After the system's implementation, fewer cases of vandalism and theft of company property were reported. Managers and security personnel could easily track the movement of workers within the building. Therefore, the project's implementation fulfilled this assumption of improving company security.

Fourth, a post-assessment of the project indicated the company significantly reduced costs associated with attendance management. The company eliminated expenses relating to purchasing stationery, such as pens and papers. Equally, it was able to reduce costs relating to human resources because the biometrics system introduced an electronic method of marking attendance. A simple detection that a fingerprint matches one in the database could automatically mark an individual present in the system and the information was stored in the server for later retrieval. Equally, reports could be generated automatically, rendering the use of the workforce unnecessary.

Therefore, this system fulfilled the assumption that attendance management costs could be significantly reduced.

Phases of the Project

The project had four phases: initiation, planning, execution, and closure.

1. Initiation Phase

The initiation phase was the first step in the project's development cycle. During this phase, the team and the project manager developed a business case for the project. They conducted feasibility studies and established a project charter, ensuring that all stakeholders understood the project's significance. The project team met with all stakeholders to listen and collect their views, they conducted feasibility tests, and assessed the current system to establish loopholes that required improvements.

2. Planning Phase

This phase entailed plenty of planning. During it, the project team prepared a detailed proposal that entailed all aspects of the project. The team considered the specifics of the project and completed the following aspects:

- Project plan

- Financial plan

- Resource plan

- Risk plan

- Quality plan

- Communications plan

- Acceptance plan

- Procurement plan

3. Execution Phase

During the execution phase, the experts built the deliverables, and the project manager monitored and controlled all tasks. In this phase, the project manager went through all management procedures step-by-step and performed a risk assessment to pinpoint risks to anticipate throughout the project's life cycle. The project manager performed:

- Time management

- Quality management

- Cost management

- Risk management

- Change management

- Procurement management

- Communications management

- Acceptance management

4. Closure Phase

In the closing phase, the project team tied up all loose ends and conducted project closure activities. The project manager, in conjunction with the project team, then reviewed the project completion. They measured the objectives and the benefits, compared the project's actual cost to the budget, and evaluated the final deliverables. Subsequently, they identified its key milestones and achievements, conducted the documentation process to guide future projects, and communicated the accomplishments to the executives and other stakeholders.

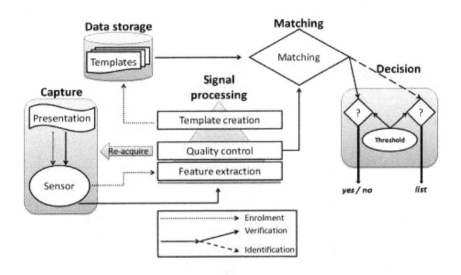

Figure 2: **Architecture of a Biometrics System**

Deviations in the Project Time Frame

Initially, the project team had stipulated that the project would take 21 days to complete. However, this time frame became exceedingly difficult due to several hiccups during its implementation. The initiation phase, which was expected to take a day, ended up taking seven days, thus disrupting all other phases of the project. The planning and design phase, which only took a week to complete, took seven extra days due to delays in preparing a detailed proposal. The team failed to execute the project within the stipulated timeframe of seven days due to an extended formative evaluation process where a few errors had to be corrected regularly. This phase required two extra weeks to be completed fully. In total, the project took an extra month to complete all of the deliverables.

Table 1: **Initial Anticipated Timelines**

Phase	Time frame	Week 1 (May 1–May 8)	Week 2 (May 8—May 15)	Week 3 (May 15 – May 22)	Week 4 (May 26)
Initiation					
Planning and Design		▓	▓		
Development and Installation			▓		
Monitoring and Evaluation				▓	
Closure					▓

Table 2: Actual Project Timeline

Phase	Timeframe	Week 1 (May 1–May 8)	Week 2 (May 8—May 15)	Week 3 (May 15 – May 22)	Week 4 (May 22 – May 29)	Week 5 (May 29 – June 05)	Week 6 (June 05 – June 12)	Week 7 (June 12 – June 19)	Week 8 (June 19 – June 26)
Initiation		▉							
Planning and Design			▉	▉					
Development and Installation					▉	▉	▉		
Monitoring and Evaluation								▉	
Closure									▉

Project dependencies

Project design can only start when its charter has been approved. The accomplishment of network infrastructure can only finish when the installation of server equipment has been complete. The software integration can only begin once the network configuration is finished. Training can start if all technical work is finished.

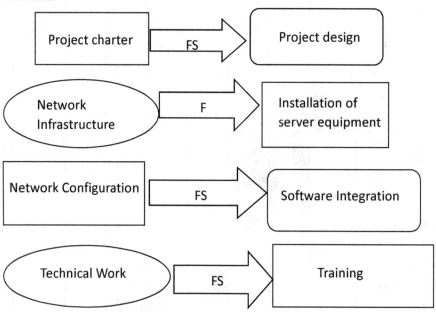

FS— Finish to Start

FF— Finish to Finish

Resource Requirements

The fingerprint biometrics system is a laptop-based application with a real-time fingerprint reader connection. The hardware and software requirements of the system are as follows.

1. Hardware Requirements

- Personal Computer (Laptop or Desktop) with the following specifications:

- 2 GB of RAM (4GB and above is recommended for maximum performance).

- Processor with a minimum clock speed of 1.0GHZ (Any recent 64-bit processor is recommended).

- Minimum of 25GB available disk space.

- USB 2.0 ports and Serial port.

- Digital Personal Fingerprint Reader

2. Software Requirements

- Microsoft NET Framework 3.5 SP1 (Service Pack One) or newer.

- Windows Operating System (Windows XP and newer).

- Microsoft Visual Studio 2012.

- Microsoft Access Database 2003 or newer versions.

- Digital Personal Fingerprint Software Development Kit.

3. Integrated Development Environment

The fingerprint biometrics system was implemented using Microsoft Visual Studio. To simplify deployment and synchronization, the database was developed using a Microsoft Access Database.

4. **Tracer Plus Desktop**

5. **Digital Person Software Development Kit**

This software development kit helped developers and integrators swiftly add the power of fingerprint-based authentication to their Microsoft® Windows-based applications.

Important Project Milestones

The following were the important project milestones:

- Milestone 1: Approval of project charter

- Milestone 2: Completion of project design

- Milestone 3: Submission and approval of the plan

- Milestone 4: Accomplishment of network infrastructure

- Milestone 5: Installation of server equipment

- Milestone 6: Completion of network configuration

- Milestone 7: Integration of software

- Milestone 8: 95% of all technical work has been completed

- Milestone 9: First training session is scheduled

- Milestone 10: Training has been complete, feedback received, and finetuning made

- Milestone 11: Key stakeholders accept the deliverables and close the project

Software and Hardware Deliverables

A. Assessing infrastructure deliverable

    i.   Checking the current topography within the organization's deliverable

    ii.   Assessing all the required resources needed to implement the system deliverable

    iii.   Purchasing the tools necessary to implement the system deliverable

B. Installation of the biometrics fingerprint deliverable

    i.   Installing the individual hardware components deliverable

    ii.   Implementing the software components deliverable

    iii.   Linking the biometrics data to individual employees' deliverables

    iv.   Connecting the system to a database deliverable

Deliverables Document and Submit

A. Assessing infrastructure deliverable

    ii.   Checking the current topography within the organization's deliverable

    iii.   Evaluating all the required resources needed to implement the system deliverable

iv.   Purchasing the tools necessary to implement the system deliverable

B. Assessing ergonomic issues or human factors deliverable

i.   Collecting stakeholder views deliverable

ii.   Evaluating stakeholder views on deliverable

iii.   Summarizing stakeholder views deliverable

iv.   Noting the major requirements of stakeholders; deliverable

C. Reviewing the feedback system deliverable

i.   Collecting the loopholes of the current system deliverable

ii.   Reviewing and identifying major upgrades deliverable

D. Checking population dynamics deliverable

i.   Collecting the current organization's demographic data

ii.   Collecting fingerprint biometrics of current staff

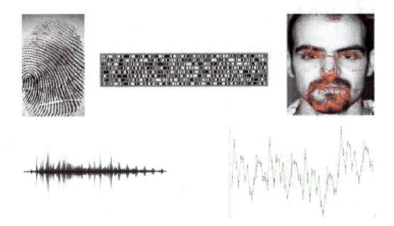

Figure 3: **Biometric Sample Template**

E. Installation of the biometrics fingerprint deliverable

i.   Installing the individual hardware components deliverable

ii.   Implementing the software components deliverable

iii. Linking the biometrics data to individual employees' deliverables

iv.   Connecting the system to a database deliverable

Part D: Revisions Made During Implementation of the Project Due to Formative Evaluation

Errors Encountered

During the project's formative evaluation, various biometric errors were discovered during authentication. These errors included false nonmatch or false reject rate, failure to capture, and false match or false accept rate. These errors have an undesirable impact on providing an effective service delivery to the right individuals. Equally, they can lead to the system wrongly denying services to the rightful beneficiaries. The project team detected various reasons why these errors occurred. Notably, this uneven distribution was caused by the difficulty of capturing some biometric features or with some having minimal unique characteristics. In the fingerprint biometric system, some fingerprints had:

- dry skin (for instance, due to cold weather),

- worn friction ridges (for instance, due to manual labor), and

- skin disease.

Some that lacked unique characteristics were because:

- of failure to present the appropriate area of the finger, for instance. only screening the top of fingers with less characteristic structures,

- few minutiae in the ridge pattern.

Revisions Made

The first revision made by the experts during the implementation of the system was to control for the difficulty of capturing biometrics. For this, a few unique characteristics were applied to improve the calculated biometric performance by selectively removing the worst samples from the database. The risk of manipulation was avoided by self-evaluating the system rather than depending on fabricated numbers. Further, the company hired independent entities that could be trusted to conduct the evaluations, but the conditions and test environments were determined by someone else. Second, these errors led to conducting pilot tests early on before project completion. Third, a backup system or an alternative was introduced at the point of service to ensure services were not denied to anyone.

Lastly, the experts recommended that in future, the A&I company should consider using a multimodal biometrics system. This system ensures that errors during authentication are reduced. It serves those individuals well whose fingerprints lack distinctive features or have friction due to manual labor. Multimodal biometrics reduces the rate of errors because when fingerprints fail, the iris is tried instead. When the iris fails to work, the face is tried. Although this introduces a new cost aspect to the multiple sensors needed, it improves the level of identity assurance at the point of service.

Part E: Summative Evaluation

Access Control

A post-assessment of the project indicated that the introduction of the system improved security within the organization. One of the problems the company experienced before was vandalism and theft of property, with difficulty in tracking responsible individuals. One of the stakeholders' primary goals was that the system could introduce a way of tracking individuals when they enter any room within the building. Consequently, this biometrics system simplified the process of tracking potential suspects of vandalism and theft. After the system's implementation, fewer cases of vandalism and theft of company property were reported. Managers and security personnel could easily track the movement of workers within the building. Therefore, the project's implementation fulfilled this goal of improving company security through enhanced access control.

Reduced Time Theft

A post-assessment of the project indicated that time theft had significantly reduced in the organization after the implementation. The number of workers who could skip job duties for personal business had reduced significantly. Those who neglected job duties had to do so after prior permission from the human resource office. This reduction had positive implications for the organization. First, the time taken for project delivery was reduced immensely. Before this system's implementation, the time for customer project delivery was lengthy because of non-corporations among team members due to consistent absenteeism. This lengthy delivery time was a turn-off for many clients who could later switch to rival companies. However, after this fingerprint biometrics system was implemented, workers were often present and punctual for their duties. This reduction in time ensured that the organization retained its current customers and even helped to attract new clients.

Equally, the organization managed to get value out of its human resource costs due to the elimination of time theft. The time, labor, and output from individual workers reflected the amount on the individual's payslip. When workers work less than the stipulated time, they are stealing from the organization because they are not giving value to the money they are paid. This time theft is a big blow for an organization because it will scare away customers and lead to losses being made. Therefore, the project fulfilled this first assumption.

Seamless management Process by Eradicating Paperwork

A post-assessment of the project showed that the system created a seamless management process because the use of paperwork for attendance management had been reduced significantly within the organization. One of the major problems before the project was that the organization spent significant resources on the paper-oriented system for attendance management. This meant it incurred extra costs for labor because more workers were needed to help take and review attendance. Reviewing all the paper-based records was problematic because of the volume of data and the tendency for distortion, theft, and loss. Notably, this complicated the matrix in performance appraisals for individual workers and the generation of pay slips based on worker attendance. Equally, workers wasted a lot of time each morning and evening by signing in and out. Before the system's installation, stakeholders assumed it could eradicate these problems and create a seamless organizational management process. Managers could easily conduct performance appraisals using electronically generated reports, review worker attendance without major hurdles, and reduce human resource costs on attendance management. Therefore, this system's installation fulfilled the goal and could introduce a seamless management process.

Reduced Attendance Management Costs

Fourth, a post-assessment of the project indicated that the company significantly reduced costs associated with attendance management. It was able to eliminate expenses relating to purchasing stationery, such as pens and papers. Equally, the company was able to reduce the costs associated with human resources because the biometrics system introduced an electronic method of marking attendance. A simple detection that a fingerprint matches one in the database could automatically mark an individual present in the system and then store the information in the server for later retrieval. Equally, reports could be generated automatically, rendering the workforce unnecessary. Therefore, this system fulfilled the goal of reducing costs associated with attendance management.

Part F: Reporting

Stakeholders

The primary stakeholders of this project included the executives and employees of the organization. These individuals were part and parcel of the implementation process because they were involved in giving their views during the project initiation phase. Equally, the project team kept them regularly updated on developments during the implementation process. These constant updates were necessary for ensuring that the project team was on track and working on the deliverables that the stakeholders desired. These progress updates contained general information such as current and previous tasks, the time they took, project manager comments, timeline completion status, overall project completion and budget spent, project risks and issues, and action items. The updates were mostly disseminated via email and conference presentations. These weekly status reports were the easiest way of keeping the stakeholders and the team informed. They were essential as well for managing expectations as the project progressed.

Similarly, as the project progressed, the project manager shared its health report with the stakeholders. This report ensured that stakeholders had a snapshot of the project status. The project manager used it to inform about action items that were on schedule and those that were running behind. This information was often shared on-screen during meetings with stakeholders and the team.

At the end of the project, all employees, including the executives, were trained on how to use the security and attendance management systems. During the training and presentations, the project manager shared general on-screen information on how the system operates, key security safeguards, and maintenance programs. Ultimately, they shared a detailed report with the executives via email, highlighting the goals, objectives, deliverables, costs, security measures, maintenance programs, and guidelines for future implementations. A physical report would then be made available where the stakeholders could sign on all the deliverables before the project was officially closed.

Project Sponsors

The project sponsors received the final project report. The project manager delivered both a soft copy and a hard copy to the project sponsor. The soft copy report was disseminated via email, while the hard copy was delivered physically for approval of all deliverables and signing off the project. This closure report recorded the final project sign-off from the sponsor, evaluated its accomplishments, and initiated some activities. The report identified and shared best practices for future projects, pinpointed and assigned items that required addressing (future initiatives, issues, future projects), appropriately closed current contracts, released project resources, provided final information to stakeholders, and ensured a proper operational transition.

The input into the project closure report emanated from various sources that included feedback from stakeholders and those involved in the project as per the defined roles in the project charter (objectives, scope, and success criteria), project documentation, risks, project management plan, issues, approved project changes, statement of work and contracts, stakeholder register, and lessons learned. Feedback was in the form of interviews, surveys, and group meetings.

Part G: Post-Implementation Support

Resources Needed for Post-implementation Support

A fingerprint biometric system is subject to wear and tear like any other system. The more the hardware and software resources of this system are used, the more they become worn out and require maintenance or updates. Software is prone to become obsolete after some time, threatening to expose the system to a cyberattack that may expose users' private information. This exposure is detrimental to both the organization and the future use of biometrics because privacy and data protection is an important industry regulation that, when breached, has far-reaching consequences. Consequently, this necessitates constant software updates and license renewal to ensure that no data breach occurs. Newer technology, especially hardware that streamlines operations, is discovered daily. Subsequently, this increases the need to regularly upgrade the hardware components to enjoy the benefits of newer technology and discard dysfunctional parts that can deny users services at the point of service. Resources such as fingerprint sensors require constant maintenance, and spare parts should be readily available to replace any dysfunctional parts. Therefore, software and hardware resources are still required in post-implementation support.

Human resources are equally important for the post-implementation support of the system. If the organization lacks skilled personnel to serve as an administrator and technical maintenance staff for the system, it may opt to train some of its current employees or hire new ones with the requisite skills. Alternatively, it can secure these services from third-party providers who can regularly perform system technical checks. Therefore, human resources are required for post-implementation support.

Detailed Plan for Short-term and Long-term Maintenance

The organization should ensure that it has plans for short-term and long-term maintenance of the system. The short-term maintenance planning approach applies data from different systems, such as maintenance planning, condition monitoring, and management systems for spare parts and equipment. The organization can use a decision support tool to select maintenance activities. The long-term maintenance planning approach entails applying condition monitoring data for assessing the optimal replacement policy.

1. **Short-term Maintenance Plan**

Regular maintenance schedules should be generated weekly. These schedules should be planned seven days in advance to allow easier planning and avoid inconveniences. All of these schedules need to be prioritized according to urgency and importance. They should be planned into the monthly workload of the maintenance crews to ensure a well-balanced selection of work for each employee without one becoming overloaded.

Weekly Maintenance Schedule				Date
Action Item	Week 1	Week 2	Week 3	Week 4
Software maintenance	Employee Name	Employee Name	Employee Name	Employee Name
Data security	Employee Name	Employee Name	Employee Name	Employee Name
Network configuration	Employee Name	Employee Name	Employee Name	Employee Name
Hardware maintenance	Employee Name	Employee Name	Employee Name	Employee Name

2. Long-term Maintenance Plan

The organization can equally run long-term maintenance schedules on the system. Creating a long-term maintenance plan within a company is important because it helps to show how many hours preventive maintenance work will take in the organization within 18-24 months. It accords the organization visibility of the required time for this preventive maintenance so they can prepare proactively for any sensitive work that may take place during that period.

Week Commencing		04-July 05	11-July 05	18-July05	25-July 05
Operating Unit	Craft	Week 1	Week 2	Week 3	Week 4
Area 1	Mech	2 M	3M	2M	3M
	Soft	1 S	2S	1S	2S
	Inst	0.5 I	1I	0.5I	1I
Area 2	Mech	2 M	3M	2M	3M
	Soft	1 S	2S	1S	2S
	Inst	0.5 I	1I	0.5I	1I
Area 3	Mech	2 M	3M	2M	3M
	Soft	1 S	2S	1S	2S
	Inst	0.5 I	1I	0.5I	1I
Area 4	Mech	2 M	3M	2M	3M
	Soft	1 S	2S	1S	2S
	Inst	0.5 I	1I	0.5I	1I
	Mech	16M	12M	16M	12M

Weekly Totals	Soft	4S	8S	4S	8S
	Inst	2I	4I	2I	4I
Monthly Totals	Mech	40 Mechanical			
	Soft	24 Software			
	Inst	12 Instrumentation			

Part H: Post-Implementation Project Summary

Deliverables Included In Documentation

The following deliverables will be included in the final documentation:

A. Reviewing the feedback system deliverable

I.   Collecting the current system's loopholes deliverable — including major discoveries identified

II. Reviewing and noting major upgrades deliverable

B. Checking population dynamics deliverable

I.   Collecting the current organization's demographic data

II. Collecting fingerprint biometrics of current staff

C. Installation of the biometrics fingerprint deliverable

I.   Installing the individual hardware components deliverable — Pictures and videos to show the installation process

II. Implementing the software components deliverable — Pictures and videos to show the installation process

III. Linking the biometrics data to individual employees' deliverables — Pictures and videos to show the installation process

IV. Connecting the system to a database deliverable — Pictures and videos to show the installation process

V.   Stakeholder feedback deliverable — Some of the review samples of major stakeholders with identities were kept anonymous

Criteria for Evaluating Each Outcome

Each outcome will have varying approaches to evaluation. Key performance metrics include data quality, security, and usability.

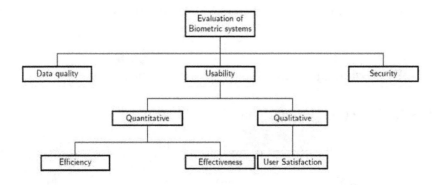

Figure 4: **Measuring the Outcomes of a Biometrics System**

Data Quality

Quality information is important because it assists experts in removing poor samples during the enrollment phase or rejecting it during the verification phase. According to the International Organization for Standardization (ISO), experts divide the quality evaluation of raw biometric data into three points:

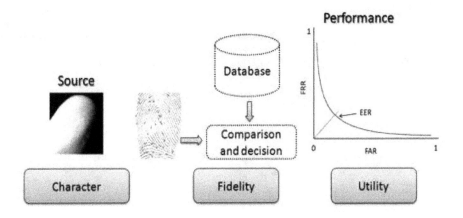

Figure 5: **Quality Assessment of Biometrics**

- Character: it is the quality of the person's physical features.

- Fidelity: the level of similarity between a biometric sample and its source (El-Abed & Charrier, 2012).

- Utility: the effect a single biometric sample can have on the general performance of a biometrics system.

This outcome was measured using various fingerprint quality assessment algorithms.

Usability

As per ISO definitions, usability is "The extent to which specified users can use a product to achieve specified goals with effectiveness, efficiency, and satisfaction in a specified context of use" (El-Abed & Charrier, 2012).

- Efficiency implies that the user can finish tasks seamlessly and on time. It was measured as task time.

- Effectiveness implies that the users can accomplish desired tasks effortlessly. The project team measured this aspect using common metrics such as failure-to-enroll-rate (FTE) and completion rate.

- User satisfaction assesses how the end-user is contented and accepts the system. The project team measured this outcome using various properties such as the level of trust and ease of use.

Security

The attack tree techniques provided a structure for the security analysis of networks, protocols, and applications. Equally, an online platform (Security Eva Bio) was used to assess the security of the biometrics system (El-Abed & Charrier, 2012). The platform applies a quantitative-based approach to security evaluation on the aspect of risk factors to simplify the evaluation comparison of biometrics systems. Further, the platforms provide a database where other researchers can use documented and common vulnerabilities and threats of biometric systems to quantify their developed systems quantitatively and qualitatively.

Justification of Differences Between Actual and Proposed Outcomes

During the implementation of the system, experts made slight modifications to the system. These minor adjustments are reflected in the actual outcomes from the proposed outcomes. For instance, the project team discovered it was essential to introduce an alternative at the point of service to eliminate the possibility of users being prevented from accessing services due to system errors. Therefore, adjustments created some differences in the outcome.

Equally, the usability and security of the system slightly failed to meet the expected level because the organization did not purchase software that meets ISO standards. However, this shortcoming can be rectified in future long-term maintenance plans.

Lessons Learned from the Process of Completing the Project

The major lessons learned in the completion of this project is that the success and failures of large-scale biometric systems depend on good project management and the outlining of goals, matching biometric competencies with primary needs and the working environment, and a comprehensive risk and threat assessment of the system (National Research Council & Whither Biometrics Committee, 2010). Common contributors to failures include:

- Insensitiveness to stakeholder requirements and perceptions

- Unsuitable technology choice

- Insufficient support infrastructure and processes

- Presumption of a non-existent problem

- Lack of a viable business case

- Poor comprehension of population issues, such as variability among the authenticated

References

Adewole, K. S., Abdulsalam, S. O., Babatunde, R. S., Shittu, T. M., & Oloyede, M. O. (2014). Development of fingerprint biometric attendance system for non-academic staff in a tertiary institution. *Development*, *5*(2), 62-70.

El-Abed, M., & Charrier, C. (2012). Evaluation of biometric systems.

National Research Council, & Whither Biometrics Committee. (2010). Biometric recognition: Challenges and opportunities.

McKenna, S., & Sarage, J. (2015, June). Biometric analytics cost estimating. In *Proc. ICEAA* (pp. 1–9).

Olagunju, M., Adeniyi, A. E., & Oladele, T. O. (2018). Staff attendance monitoring system using fingerprint biometrics. *International Journal of Computer Applications*, *179*(21), 8-15.